Newmans'
Medical Laboratory
Assistants
Study Guide

Dear Graduate:

This study guide will provide information about the medical laboratory assistant as a specialized area of clinical laboratory practice. The role of a CMLA has expanded, thus, creating the need to replace on-the-job training with structured training programs, which, in turn, has lead to certification. The reader can use this booklet as a study guide for the Certified Medical Laboratory Assistant exam. As such, this is a supplement for a review and it is **not meant to replace training textbooks and/or lecture notes.**

TABLE OF CONTENTS

The Medical Laboratory

Rapid scientific advances in diagnostic and treatment instrumentation within health care settings have led to numerous changes within the clinical laboratory. Almost every patient who is admitted to a hospital, or seen in a physician's office, also becomes a patient in the clinical laboratory. In the past few years, rapid advances in clinical laboratory automation and procedures have spurred tremendous growth in the clinical laboratory, at the rate that has exceeded hospital growth. Sophisticated technology and automation have added new dimensions to the diagnosis and treatment of disease.

The clinical laboratory department is composed of two major areas. In the clinical pathology area, blood and other types of body fluids and tissues, such as urine, cerebrospinal fluid (CSF), biopsy specimens, and gastric secretions, are analyzed. The area of anatomic pathology is concerned with autopsy and cytological examinations, as well as surgical pathology procedures. The clinical laboratory primarily offers patient services, but it also involved in research development and teaching in order to maintain the high quality of laboratory services.

Clinical Laboratory Sections

Hematology Section

This is the section where the formed elements of the blood are studied by enumerating and classifying the red blood cells, white blood cells, and platelets. By studying and examining the cells, disorders and infections are detected and treatment instituted or monitored.

Whole blood is the most common specimen analyzed and usually collected in lavender-top tube containing the anti-coagulant EDTA.

The coagulation section is usually a part of hematology. But in large laboratories they are separated. This is the area where hemostasis is evaluated. Plasma is usually the specimen analyzed drawn from blood collected.

Chemistry Section

The most automated section in the laboratory. This section is divided into several areas:

- Electrophoresis – analyzes chemical components of blood such as hemoglobin and serum, urine and cerebrospinal fluid, based on the differences in electrical charge.

- Toxicology - analyzes plasma levels of drugs and poisons.

- Immunochemistry – This section uses techniques such as radio immunoassay (RIA) and enzyme immunoassay to detect and measure substances such as hormones, enzymes, and drugs.

Some tests in the chemistry section are ordered by profiles, which are groups of test ordered by a physician to evaluate the status of an organ, body system or general health of the patient.

> Examples of these profiles are:
> ◊ Liver profile: tests may include ALP, AST, ALT, GGT and Bilirubin
> ◊ Coronary risk profile: tests may include Cholesterol, Triglycerides, HDL, LDL

Blood Bank Section

This is the section where blood is collected, stored and prepared for transfusion. Strict adherence to procedures for patient identification and specimen handling is a must to ensure patient safety.

Blood collected may be separated into components: packed cells, platelets, fresh frozen plasma, and cryoprecipitate.

Serology (Immunology) Section

Performs tests to evaluate the patient's immune response through the production of antibodies. This section uses serum to analyze presence of antibodies to bacteria, viruses, fungi, parasites and antibodies against the body's own substances (autoimmunity).

Microbiology Section

This section is responsible for the detection of pathogenic microorganisms in patient samples and for the hospital infection control.

The primary test performed is culture and sensitivity (C&S). It is used to detect and identify microorganisms and to determine the most effective antibiotic therapy. Results are usually available within 24 to 48 hours; but cultures for tuberculosis and fungi require several weeks.

One instance when culture and sensitivity is used is to diagnose the cause of a patient's fever of unknown origin (FUO).

Quality Assurance/Quality Control

Clinical laboratory personnel performed quality work and provided quality results long before the terms quality assurance and quality control became popular. Quality assurance (QA) refers to a set of policies and procedures that are followed to ensure that every test performed is valid. These policies assure the patient that the results of the test performed are accurate and can safely be used to guide treatment. Quality control (QC) is the process by which final results are validated and variations are quantified. The principles of quality are rooted in both fact and perception. "Doing the right thing, the right way, the first time, and on time" are all aspects of quality. These aspects of quality, when applied to the heath care setting, are often driven by outside agencies, whether government-sponsored (e.g., the Food and Drug Administration [FDA]) or voluntary (e.g., accrediting agencies, such as the Joint Commission on the Accreditation of Health Care Organizations [JCAHO]).

Quality Control: To ensure accurate and reliable test results, each laboratory must establish and follow written quality control procedures that monitor and evaluate the quality of each testing process. These include developing a laboratory procedure manual, following the manufacturer's instructions for each product; performing and documenting calibration procedures at least every 6 months and two levels of controls daily; performing and documenting actions taken when problems or errors are identified; and documenting all quality control activities.

Safety in the Laboratory

Safety rules are usually based on common sense. Most accidents occur when these rules are neglected, overlooked or ignored. Accidents generally occur when safety is compromised because of haste and secondary shortcuts. These shortcuts can lead to personal injury and equipment damage. When an accident occurs, it must be reported to your supervisor immediately. Trying to cover up the incident can lead to serious, even disastrous results.

The U.S. Public Health Service's Centers for Disease Control (CDC) did considerable research on personal and environmental laboratory safety. From this research, a recommended set of safety regulations for the workplace, known as "Standard Precautions" were established, and the Occupational Safety and Health Administration (OSHA) was given the task of enforcing these precautions. According to the Standard Precautions regulations, all human blood and certain human body fluids are treated as if they are known to be infectious for human immunodeficiency virus (HIV), hepatitis B virus (HBV), and other blood-borne pathogens. The OSHA-organized Standards must be observed when dealing with all patients, regardless of their known or suspected diagnosis.

The major areas of greatest concern in the laboratory are divided into three classifications that are described as "laboratory hazards." These hazard areas are physical, chemical, and biological.

Laboratory Hazards

Physical Hazards
Electrical Safety Regulations
- ❖ Use only ground plugs that have been approved by Underwriters' Laboratory (UL).
- ❖ Never use extension cords.
- ❖ Avoid electrical circuit overloading.
- ❖ Inspect all cords and plugs periodically for damage.
- ❖ Use a surge protector on all sensitive electronic devices.
- ❖ Before servicing, UNPLUG the device from the electrical outlet.
- ❖ Use signs and/or labels to indicate high voltage or electrical hazards.

Chemical Hazards
Chemical Safety Regulations
- ❖ If the skin or eyes come in contact with any chemicals, immediately wash the area with water for at least 5 minutes.
- ❖ Store flammable or volatile chemicals in a well-ventilated area.
- ❖ After use, immediately recap all bottles containing toxic substances.
- ❖ Label all chemicals with the required MSDS information.

Biological Hazards

Biological Safety Regulations
- ❖ Disinfect the laboratory work area before and after each use when dealing with biologicals.
- ❖ Never draw a specimen through a pipette by mouth. This technique is not permitted in the laboratory.
- ❖ Always wear gloves.
- ❖ Sterilize specimens and any other contaminated materials and/or dispose of them through incineration.
- ❖ Wash hands thoroughly before and after every procedure.

Infection Control

The chain of infection requires a continuous link among three elements:

Source -------------------------- Transmission --------------------- Susceptible host

Portal	Portal
of	of
Exit	Entry

The source refers to the location of the pathogenic organism. Transmission is the method by which the microorganism is transferred to the susceptible host, which could be the patient, or for that matter, any person.

The microorganism must have a means to get out of the source (portal of exit) and a means to get inside the susceptible host (portal of entry) to complete the chain of infection. The goal of biologic safety is to prevent this completion of the chain. Specific safety practices are directed toward each component with the goal of breaking it.

An infection contracted by the patient during hospitalization is called "Nosocomial Infection." Health care providers, not following instituted control practices, cause most of these infections. Hand contact is the most common method of transmission, which is why hand washing is the most important means of prevention.

Isolation Precautions

For many years, the CDC recommended universal precautions, which is a method of infection control that assumed that all human blood and body fluids were potentially infectious. The CDC issued a revised guidelines consisting of two tiers or levels of precautions: Standard Precautions and Transmission-Based Precautions.

Standard Precautions
This is an infection control method designed to prevent direct contact with blood and other body fluids and tissues by using barrier protection and work control practices. Under the standard precautions, all patients are presumed to be infective for blood-borne pathogens. Infection control practices to be used with all patients. These replace universal precautions and body substance isolation. They are used when there is a possibility of contact with any of the following:

◊ Blood
◊ All body fluids, secretions, and excretions (except sweat), regardless of whether or not they contain visible blood
◊ Nonintact skin
◊ Mucous membranes designed to reduce the risk of transmission of microorganisms from both
◊ Recognized and unrecognized sources of infections.

The standard precautions are:
- Wear gloves when collecting and handling blood, body fluids, or tissue specimen.
- Wear face shields when there is a danger for splashing on mucous membranes.
- Dispose of all needles and sharp objects in puncture-proof containers without recapping.

Transmission- Based Precautions the second tier of precautions and are to be used when the patient is known or suspected of being infected with contagious disease. They are to be used in addition to standard precautions. All types of isolation are condensed into three categories:

Contact precautions: are designed to reduce the risk of transmission of microorganisms by direct or indirect contact. Direct-contact transmission involves skin-to-skin contact and physical transfer of microorganisms to a susceptible host from an infected or colonized person. Indirect-contact transmission involves contact with a contaminated intermediate object in the patient's environment

Airborne precautions: are designed to reduce the risk of airborne transmission of infectious agents. Microorganisms carried in this manner can be dispersed widely by air currents and may become inhaled by or deposited on a susceptible host within the same room or over a longer distance from the source patient. Special air handling and ventilation are required to prevent airborne transmission.

Droplet precautions: are designed to reduce the risk of droplet transmission of infectious agents. Droplet transmission involves contact with the conjunctivae or the mucous membranes of the nose or mouth of a susceptible person with large-particle droplets generated from the source person primarily during coughing, sneezing, or talking. Because droplets generally travel only short distances, usually three feet or less, and do not remain suspended in the air, special air handling and ventilation are not required.

Hepatitis and Acquired Immunodeficiency Syndrome (AIDS)

Hepatitis and acquired immunodeficiency syndrome (AIDS) represent constant threats to the health and safety of laboratory personnel. Both of these potentially deadly diseases are transmitted as a result of exposure to blood and body fluids from an infected individual. The infectious process is similar for both of these diseases.

Hepatitis A

Type A virus causes Hepatitis A (HAV) previously called "infectious hepatitis". It is usually spread by the fecal-oral route as a result of improper personal hygiene methods or consumption of contaminated foods, such as shellfish. It is a very common form of viral hepatitis; indeed, approximately 143,000 cases were reported in the United States in 1994. The vaccine was later introduced in 1995 in the United States of America, and health professionals know vaccinate all children, and travelers to certain countries. There are about 3-6 deaths per 1,000 cases of Hepatitis A.

www.cdc.gov

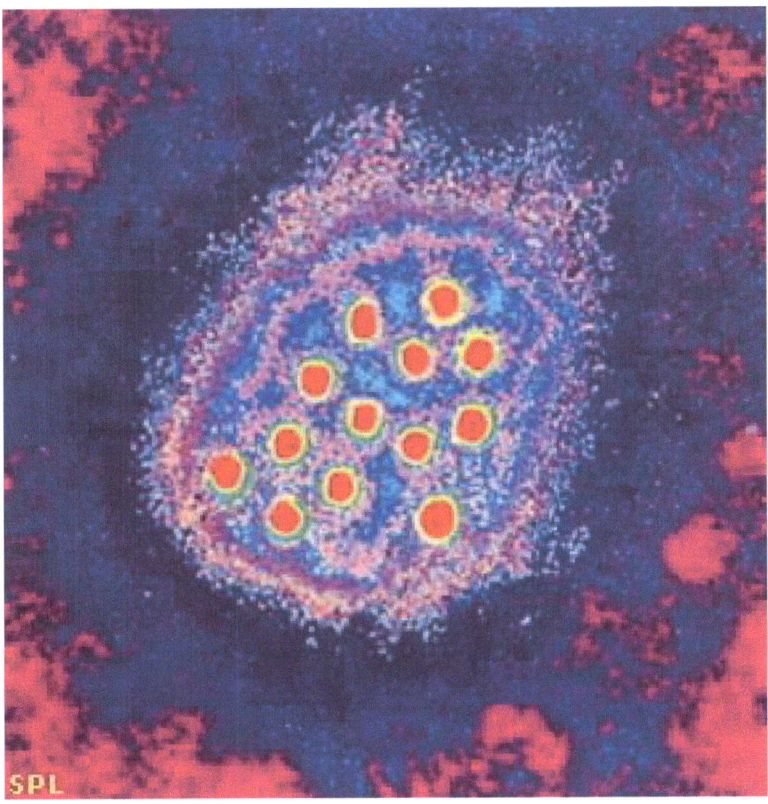

http://www.hepatitisblog.com/

Hepatitis B

Hepatitis B (HBV), previously referred to as serum hepatitis, is caused by type B virus and is acquired parenterally, through contact with blood (transfusions, hypodermic needles, dental and surgical instruments) and body fluids (tears, saliva, and semen). People who collect and process blood are considered to be at high risk for contracting HBV. Although most infected individuals recover, 10% of the population with HBV becomes chronic carriers.

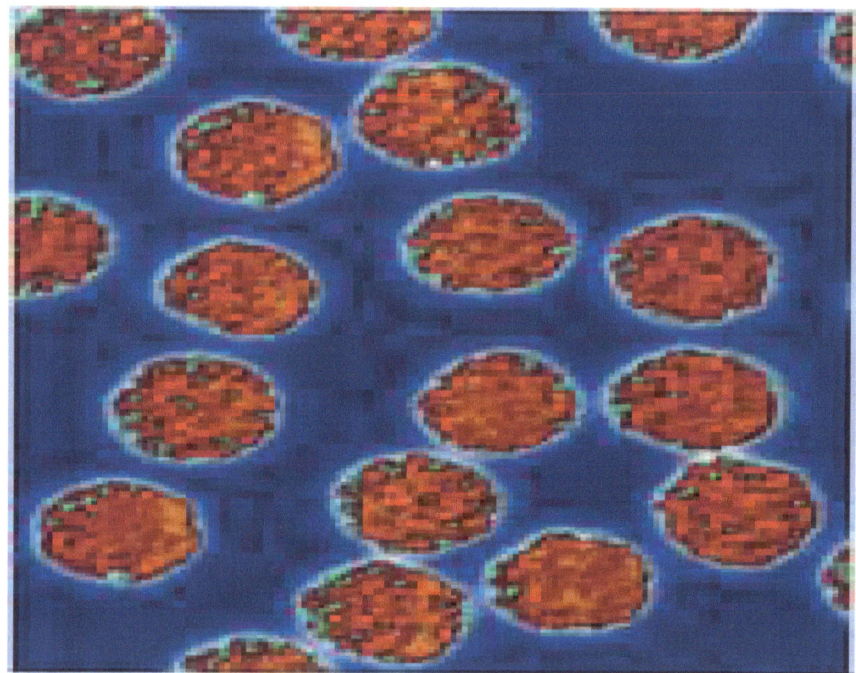

Nytimes.com

Hepatitis B Immunization

OSHA standards require physicians to offer the hepatitis B vaccination series, free of charge, to all medical office personnel who may have occupational exposure. It is a three-dose series that is believed to give immunity to the virus. The immunization must be offered within the first 10 days of employment in an occupational risk area. Laboratory personnel who do not want to be vaccinated must sign a waiver form documenting refusal, which is to be filled in the employee's OSHA record.

AIDS

AIDS is not a disease that can be casually contracted. Routine encounters with patients in the laboratory are not sources of HIV transmission. Comparing all the sexually transmitted diseases that are known to us, HIV is the most difficult to contract. Although the chances of contracting HIV are low, the serious nature of the infection makes it imperative that the laboratory assistant use precautionary procedures. AIDS is caused by a retrovirus known as HIV. Because many HIV carriers are asymptomatic and may not be aware that they have the virus, procedures to minimize the risk of exposure to blood and body fluids should be taken with *all* patients at all times. The current guidelines for diagnosing AIDS are based on the following criteria: (1) the presence of AIDS-related disorders, and (2) the T-cell count of an HIV-infected individual. The normal range for the T-cell count is 800 to 1600/cm of blood. The diagnosis of AIDS is applied to individuals with a T-cell count of 200/cm or less. The early signs of HIV occur during the first 2-4 weeks after exposure to the virus. Acute retroviral syndrome, also known as (ARS) has symptoms similar to the flu.

www.aids.gov

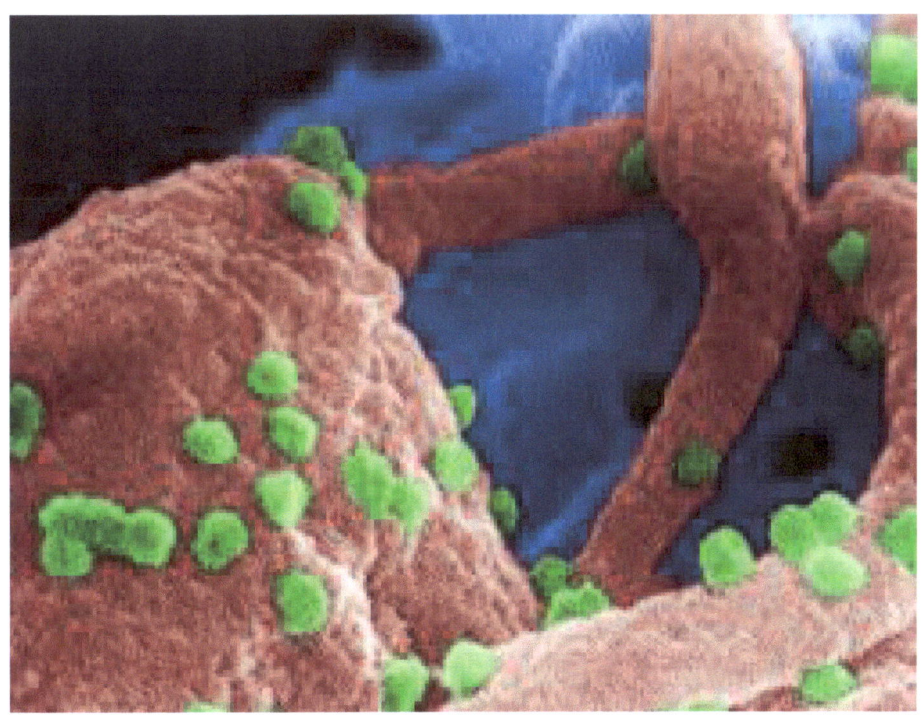

The Aids virus detects by changes in the T-cell count
www.news-medical.net

The Microscope

Every medical laboratory is equipped with a microscope. This indispensable instrument is used to view objects too small to be seen with the naked eye. The microscope is used to evaluate stained blood smears, urine sediment, and micro bacterial smears of the throat, vagina, and semen, as well as to determine blood cell counts.

Microscopes may be monocular, dual, binocular, or triocular. A monocular microscope has one eyepiece for viewing, whereas the binocular microscope has two eyepieces and the triocular microscope has three. The eyepiece or ocular is located at the top of the microscope and contains a lens that magnifies what is being seen. The ocular is attached to a barrel or tube that is connected to the microscope arm. Under the arm is the revolving nosepiece, to which are attached the DIN objectives.

Most microscopes have three or four objectives, and each has a different magnifying power. The shortest objective, which has the lowest power, is marked 10x or 16 mm. Low power is used for initial focusing and to scan urine sediment and blood smears; it may then be used to focus on an area of particular interest. Greater detail is observed with the next longest objective (high power), which is marked 40x or 4 mm. High magnifying power is used to study cells and sediment in greater detail and for manually counting red blood cells. The longest objective (oil immersion), which is marked 100x or 1.8 mm, allows for the finest focusing on an object and is used to view microorganisms, to view a Gram-stained specimen, and to obtain differential white cell counts and reticulocyte counts. If there is a fourth objective, it is labeled 4x and is used in a manner similar to the 10x objective, but only half the magnification.

When adjusting the objective to a new power, the nosepiece should be rotated using its grip, NOT THE OBJECTIVES. Otherwise, the objective may be loosened and seriously damaged by incorrect rotation.

The arm of the microscope connects the objectives and oculars to the base, which supports the microscope and contains the light source. The condenser is mounted under the stage of the microscope and can be lowered so that each field of view is filled with light rays. The iris diaphragm controls the quality and quantity of light reaching the objective by adjusting the iris leaves under the condenser. Just above the base are the focusing knobs. The course adjustment is used only with low power; the fine adjustment is used with both high power and oil immersion microscopy. The stage of the microscope holds the slide to be viewed.

The different types of microscopes are: (1) Brightfield Microscope (2) Darkfield Microscope (3) Phase-Contrast Microscope (4) Fluorescence Microscope (5) Electron Microscope.

Understanding Laboratory Measurements

Basic Units of the System

Meter

The meter is the basic unit of length. A meter is 39.37 inches long. A decimeter would be 3.94 inches long. A centimeter would be about 0.39 inches long, and a millimeter would be about 0.04 inches in length. One thousand meters would equal 1 km. Meters are often used in laboratory reports, charts, and other data requiring linear measurements. For instance, a laboratory procedure might require you to "connect flasks using 0.3 meters of rubber tubing;" this would mean that you must use 12 inches of tubing.

Liter

The liter is the basic unit of capacity or volume. This measure tells us how much space an item occupies. The standard unit for capacity in the International System is expressed in terms of multiples or decimal fractions of the cubic meter. In the laboratory, this unit is too large for everyday use; thus the cubic decimeter is used. The liter is accepted as a general designation for 1 cubic decimeter. The liter is used most frequently in the United States by the beverage industry. We are all familiar with the 2-liter soft drink container. A liter is slightly more than 1 qt and is equal to 1000 mL, or the capacity occupied by 2.2 lb of distilled water at 39.2*C.

Gram

The gram is the basic unit of weight or mass. A measure of weight tells how heavy an item is. One thousand cubic centimeters, the equivalent of 1 cubic decimeter, have the capacity of 1 L and weigh 1000 gm or 1 kg. The kilogram is the standard unit of weight and is equivalent to approximately 2.2 lb in the English system of measurement. In the clinical laboratory, the gram (0.001 kg) is used more frequently than is the kilogram. A gram is the weight of 1 cubic centimeter of distilled water at a temperature of 39.2*C.

Metric Abbreviations				
Metric Unit	Abbreviation	Meter	Liter	Gram
Meter	M (m)			
Liter	L (l)			
Gram	g or gm			
Kilo	k	km	kL	kg
Hecto	h	hm	hL	hg
Deca	da	dam	daL	dag
Deci	d	dm	mL	dg
Centi	c	cm	cL	cg
Milli	m	mm	mL	mg
Micro	u	um	uL	ug
Nano	n	nm	nL	ng

```
┌─────────────────────────────────────────────────────────────┐
│  Commonly Used Metric Prefixes                                │
│                                                               │
│  Kilo= 1000.00                    (One-thousand)              │
│  Deci= 0.1                        (one-tenth)                 │
│  Centi= 0.01                      (one-hundredth)             │
│  Milli= 0.001                     (one-thousandth)            │
│  Micro= 0.000,001                 (one millionth)             │
│  Nano= 0.000,000,001              (one billionth)             │
│                                                               │
└─────────────────────────────────────────────────────────────┘
```

Metric Measurement

Because the metric system is based on 10 and multiples of 10, 100, 1000, etc. the measurements can be expressed in decimals, and decimals can be used in the calculations. Fractions should be changed to their decimal equivalents when working with the metric system.

To express one unit of measure in terms of the next smaller unit, multiply by 1000. Therefore 2 gm equals 2000 mg and 0.5 gm equals 500 mg. To express on unit of measure in terms of the next larger unit, divide by 1000. Therefore, 2 mg equals 0.002 gm and 50 mg equals 0.05 gm.

Multiples of the metric unit is preceded by prefixes derived from Greek and Latin. The most frequently used prefixes in the clinical laboratory are: kilo (k), milli (m), deci (d), centi ©, micro (u), and nano (n).

Solutions and Dilutions

It is necessary to make dilutions in the laboratory frequently. For example, blood, serum, or plasma are diluted to produce color reactions that can be used in determining test results. When blood cell counts are done manually, it is necessary to make a dilution before these cells can be counted under the microscope.

Today, most solutions are commercially prepared and come to the laboratory in a ready-to use package. You know that a 10% bleach solution is the solution of choice in cleaning areas where there is the possibility of body fluid contamination. There will be a time that you will need to prepare a solution of certain strength from a given solution of another strength. Whenever solution preparation is required, accuracy is essential. When preparing a solution, it must be accomplished to exact specifications.

Preparing Solutions and Dilutions

Whenever a dilution is to be prepared, the formula is as follows:

$$\frac{\text{Desired strength}}{\text{Available strength}} = \frac{\text{X (amount needed)}}{\text{amount available}}$$

Therapeutic Drug Monitoring

Therapeutic Drug Monitoring (TDM) is a means by which the physician can measure the effects or levels of a drug being administered to a patient.

There are different ways by which medications can be monitored. When giving insulin, a physician can measure the blood glucose level to determine what effect the insulin is having on the body. Simply obtaining a blood pressure measurement can monitor the efficacy of medications to control blood pressure. There are also drugs whose affects can only be assessed by actual measurements of drug levels in the body. One example of such a drug is digoxin. To a certain degree, determining the heart rate can monitor the effects of digoxin. However, digoxin has a narrow therapeutic range, which means it can quickly become toxic to the patient. Thus, the aim of digoxin therapy is to give enough medication to be effective, but not enough to produce symptoms of toxicity. On the basis of pharmaceutical studies, therapeutic ranges for different drugs have been established which gives the physician guidelines for prescribing medications. The "fine tuning" required to prescribe the exact effective amount of medication can be accomplished by determining the drug level.

For certain drugs, it is very important to collect a blood sample for TDM at specific times either before or after administering the drug. Otherwise the test results will be useless and unreliable, or even worse; they may prompt the physician to initiate inappropriate treatment. Digoxin is a good example of one such drug. Shortly after a dose of digoxin (the period depends on the method of administration), the blood levels are high, but the patient shows no sign of intoxication. After this, the drug taken up by the target tissue (the heart) and blood levels become quite low. At that point, the blood levels are in equilibrium with the levels in the target tissue, correlating with clinical status.

This test is used to monitor the blood levels of certain medication to ensure patient safety and also maintain a plasma level. Blood is drawn to coincide with the trough (lowest blood level) or the peak level (highest blood level).

Trough levels are collected 30 minutes before the scheduled dose. Time for collecting peak level will vary depending on the medication, patient's metabolism, and the route of administration (I.V., I.M., or oral).

Arterial Blood Gas Studies

Arterial blood gas studies (ABGs) are valuable tools in the treatment of critically ill patients. As the name suggests, ABGs are one of the few clinical laboratory procedures performed on arterial blood. Arterial blood gas analyzers quantitate ABG components using special electrodes. ABGs help assess a patient's ventilation, oxygenation, and acid-base balance. ABGs are also used to monitor the condition of critically ill patients, to diagnose electrolyte imbalances, to monitor oxygen flow rates, and to complement other pulmonary function studies. It should be remembered that any arterial puncture should not be attempted by anyone who is not trained and licensed to perform this procedure.

References Ranges for Frequently Ordered ABG Tests

TESTS	REFERENCE RANGE
pH	7.35-7.45
O2 content	15-22 vol%
PO2	> 80 torr*
SO2	> 95% of capacity
CO2 content	25-30 nmol/L
PCO2	35-45 torr*
HCO3	24-28 mEq/L
Base excess	>3 mEq/L
Base deficit	<3 mEq/L

*torr=a unit of pressure equal to exactly 1/760 atmosphere and nearly equal to 1 mm Hg.

Infectious Mononucleosis

Infectious mononucleosis is an acute viral infection caused by the Epstein-Barr virus (EBV). It is also called "mono" and is sometimes abbreviated "IM." Infectious mononucleosis is a disease commonly found in adolescents and young adults. Because it is believed to be transmitted orally, it is sometimes called the "kissing disease." The vague, flu-like symptoms associated with infectious mononucleosis, makes diagnosis difficult, and therefore, diagnosis is usually based on the evaluation of clinical symptoms, hematologic testing, and serologic testing. Infectious mononucleosis is characterized by fever, pharyngitis, swollen lymph glands, atypical lymphocytes, splenomegaly, hepatomegaly, fatigue, weakness, and headache. Infection typically creates permanent immunity. Treatment is primarily symptomatic and usually includes bed rest.

Hematologic testing for infectious mononucleosis involves a complete blood count (CBC). Particular attention is paid to the patient's total white blood cell count (WBC) and the appearance of the patient's lymphocytes. In infectious mononucleosis, the lymphocytes may have an unusual or atypical appearance, and the patient may have an increased number of lymphocytes, termed lymphocytosis. Most patients exposed to EPV develop a heterophil antibody response. Heterophils are part of the widespread group of antibodies that are characterized by their ability to react with the surface antigens present on the red blood cells (RBCs) of different mammalian species. Most qualitative mononucleosis tests are slide tests. Slide tests yield rapid, reliable results, and are easy to perform. Specialized tests to detect the presence of EPV are available in reference laboratories and government-run public health laboratories.

Serologic test kits for infectious mononucleosis provide all the necessary reagents, materials, and controls. All that is needed is a small amount of the patient's plasma or serum. Some kits even allow testing of whole blood. Traditional tests for infectious mononucleosis, such as "Mono-test" are based on the agglutination of horse erythrocytes by the heterophil antibody present in patients with infectious mononucleosis. Because there are other antibodies present that will also react with these interfering antibodies before it is combined with the horse erythrocytes. Numerous commercial kits for testing for infectious mononucleosis are available, including Pulse Test, Monoslide, Monospot, and the Color Slide II.

Testing Procedures

Determination of ABO Group

Determination of ABO blood groups is a simple test that can easily be performed. The test detects the presence of A or B antigens on RBCs based on the presence or absence of agglutination with a known antiserum. When the antigen on the patient's RBCs corresponds to the antibody, agglutination occurs. If the corresponding antigen is not present on the RBCs, no agglutination will occur. For example, Type A blood will agglutinate in the presence of anti-A antiserum, but in the presence of anti-B antiserum. Type B blood will agglutinate in the presence of anti-B antiserum, but not in the presence of anti-A antiserum. Type O blood will not agglutinate in the presence of anti-A antiserum or anti-B antiserum, whereas Type AB blood will agglutinate in the presence of both anti-A antiserum and anti-B antiserum. The test can be performed by using either a slide test or a tube test.

Venipuncture

The basic step in performing venipuncture is to have the necessary supplies and/or equipment organized for proper collection of specimen and to ensure the patient's safety and comfort. The recommended supplies are as follows:

A. Laboratory requisition slip and pen.

B. Antiseptic –
   - Prepackaged 70% isopropyl alcohol pads are the most commonly used.
   - For collections that require more stringent infection control such as blood cultures and arterial punctures Povidone-iodine solution is commonly used.
   - For patients allergic to iodine, chlorhexidine gluconate is used.

C. Vacutainer tubes –
   - Color-coded for specific tests and available in adult and pediatric sizes.

D. Vacutainer needles-
   - These are disposable and are used only once both for single-tube draw and multidraw (more than one tube).
   - Needle sizes differ both in length and gauge. 1-inch and 1.5-inch long are routinely used.
   - The diameter of the bore of the needle is referred to as the gauge. The smaller the gauge the bigger the diameter of the needle; the bigger the gauge the smaller the diameter of the needle (i.e. 16 gauge is large bore and 23 gauge is small bore.) Needles smaller than 23 gauge are not used for drawing blood because they can cause hemolysis.

E. Needle adapters -
   - Also called the tube holder. One end has a small opening that connects the needle, and the other end has a wide opening to hold the collection tube.

F. Winged infusion sets -
   - Used for venipuncture on small veins such as those in the hand. They are also used for venipuncture in the elderly and pediatric patients.
   - The most common size is 23gauge, ½ to ¾ inch long.

G. Sterile syringes and needles -
   - 10-20 ml syringe is used when the Vacutainer method cannot be used.

H. Tourniquets –
   - Prevents the venous outflow of blood from the arm causing the veins to bulge thereby making it easier to locate the veins.
   - The most common tourniquet used is the latex strip. (Be sure to check for latex allergy). Tourniquets with Velcro and buckle closures are also available.
   - Blood pressure cuffs may also be used as tourniquet. The cuff is inflated to a pressure above the diastolic but below the systolic.

I. Chux –
   - An impermeable pad used to protect the patient's clothing and bedding.

J. Specimen labels -
   - To be placed on each tube collected after the venipuncture.

K. Gloves -
   - Must always be worn when collecting blood specimen

L. Needle disposal container –
   - Must be a clearly marked puncture-resistant biohazard disposal container.
   - **Never recap a needle without a safety device.**

Site Selection

The preferred site for venipuncture is the antecubital fossa of the upper extremities. The vein should be large enough to receive the shaft of the needle, and it should be visible or palpable after tourniquet placement.

Three major veins are located in the antecubital fossa, and they are:

A. **Median cubital vein** – the vein of choice because it is large and does not tend to move when the needle is inserted.
B. **Cephalic vein** - the second choice. It is usually more difficult to locate and has a tendency to move, however, it is often the only vein that can be palpated in the obese patient.
C. **Basilic vein** - the third choice. It is the least firmly anchored and located near the brachial artery. If the needle is inserted too deep, this artery may be punctured.

Unsuitable veins for venipuncture are:
  A. **Sclerosed veins** - These veins feel hard or cordlike. Can be caused by disease, inflammation, chemotherapy or repeated venipunctures.
  B. **Thrombotic veins**
  C. **Tortuous veins** – These are winding or crooked veins. These veins are susceptible to infection, and since blood flow is impaired, the specimen collected may produce erroneous test results.

Note: Do not draw blood from an arm with IV fluids running into it. The fluid will alter the test results. Select another site. Do not draw blood from an artificial a-v fistula site, such as those surgically implanted in dialysis patients.

Complications Associated With Phlebotomy

Hematoma:
The most common complication of phlebotomy procedure. This indicates that blood has accumulated in the tissue surrounding the vein. The two most common causes are the needle going through the vein, and/or failure to apply enough pressure on the site after needle withdrawal.

Hemoconcentration:
The increase in proportion of formed elements to plasma caused by the tourniquet being left on too long. (More than two (2) minutes)

Phlebitis:
Inflammation of a vein as a result of repeated venipuncture on that vein.

Petechiae:
These are tiny non-raised red spots that appear on the skin from rupturing of the capillaries due to the tourniquet being left on too long or too tight.

Thrombus:
This is blood clot usually a consequence of insufficient pressure applied after the withdrawal of the needle.

Thrombophlebitis:
Inflammation of a vein with formation of a clot

Septicemia:
This is systemic infection associated with the presence of pathogenic organism introduced during a venipuncture.

Trauma:
This is injury to underlying tissues caused by probing of the needle.

Factors To Consider Prior To Performing The Phlebotomy Procedure:

Fasting – some tests such as those for glucose, cholesterol, and triglycerides require that the patient abstain from eating for at least 12 hours. The phlebotomist must ascertain that the patient is indeed in a fasting state prior to the testing.

Edema –is the accumulation of fluid in the tissues. Collection from edematous tissue alters test results.

Fistula - is the permanent surgical connection between an artery and a vein. Fistulas are used for dialysis procedures and must never be used for venipunctures due to the possibility of infection.

ROUTINE VENIPUNCTURE

1) Verify the requisition for the tests.
2) Identify the patient: check the patient's ID number or have him/her state his/her name.
3) Identify yourself to the patient, explain the procedure and secure his/her consent.
4) Palpate the veins in the antecubital fossa using your index finger.
5) Gather the necessary equipment.
6) Wash hands; put on gloves.
7) Tie on the tourniquet; it should be applied 3-4 inches above the site where the venipuncture will be made. Ask the patient to make a fist or open and close his/her hand to help engorge the vein.
8) Palpate the vein while looking for the straightest point. Cleanse the area using a circular motion starting at the inside of the venipuncture site.
9) Assemble the needle and tube holder while the alcohol is drying. Uncap the needle and examine it for defects such as blunted or barbed point.
10) Hold the patient's arm, by placing four fingers under the forearm and your thumb below the antecubital area slightly pulling the skin back to anchor the vein.
11) With the bevel facing upward, insert the needle at an angle of 15-30 degrees.
12) Once the needle is inside the vein (you will feel a "give" as the vein is entered), push the collection tube into the holder to puncture the tube stopper with the back-end of the needle.
13) Fill the needed tubes, according to the order of draw.
14) While filling the last tube, release the tourniquet. Taking into consideration that the tourniquet should not be left on for more than one (1) minute.
15) Pull out collection tube from the holder.
16) Place folded gauze over the venipuncture site and withdraw the needle. Then apply pressure until bleeding stops. This is done to prevent hematoma. (The patient, if capable, can be asked to apply this pressure.)
17) Discard needle into the biohazards sharp container.

18) Label each collected specimen, writing the patient's name and ID number, the time and date of collection, and your initials.
19) Place labeled tubes inside the biohazards transport bag.
20) Before leaving, check the venipuncture site. If it is still bleeding, apply pressure for another 2 minutes. If after this time, it is still bleeding, continue to apply pressure for another 3 minutes. If bleeding persists after a total of 8 minutes of applying pressure, call for help.
21) At any point when the bleeding stops, an adhesive bandage is applied over a folded gauze square. The patient should be instructed to remove the bandage within an hour.
22) Clean up everything and dispose of waste properly.
23) Leave the patient's call light within his/her reach.
24) Remove the gloves, wash your hands, say good-bye to the patient and inform him/her that his/her physician will deliver the results.

- ✓ **Do not label the tubes prior to the venipuncture.**
- ✓ **Do not leave the patient's room before labeling the tubes.**
- ✓ **Do not dismiss an outpatient before labeling the tubes.**
- ✓ **Do not label tubes using a pencil; black ink should be used.**
- ✓ **Do not leave the patient until you checked and ensure that the bleeding has stopped.**

Failure to Obtain Blood

Most venipunctures are routine, but in some instances, complications can arise resulting in failure to obtain blood. The following are some of the common causes:

- ❖ The tube has lost its vacuum. This is maybe due to:
  - ➢ A manufacturing defect
  - ➢ Expired tube
  - ➢ A very fine crack in the tube

- ❖ Improperly positioned needle. In many instances, slight movement of the needle can correct this.
  - ➢ The bevel of the needle is resting against the wall of the vein. Slightly rotate the needle.
  - ➢ The needle is not fully in the vein. Slowly advance the needle.
  - ➢ The needle has passed through the vein. Slowly pull back on the vein.
  - ➢ The needle was missed completely. With a gloved finger gently determine the positions of the vein and the needle, and redirect the needle.
  - ➢ Collapsed vein. This maybe due to excessive pull from the vacuum tube and use of a smaller vacuum tube may remedy the situation. If it does not, remove the tourniquet, withdraw the needle, and select another vein preferably using either a syringe or butterfly.

Special Venipuncture

Some venipunctures are done using special collecting or handling procedures specific to the test being requested. Some require patient preparation such as fasting, while some needs to be collected at a specific time. Still, others may need special handling such as protection from light.

Fasting Specimens

This requires collection of blood while the patient is in the basal state, that is, the patient has fasted and refrained from strenuous exercise for 12 hours prior to the drawing.

It is the phlebotomist responsibility to verify if the patient indeed, has been fasting for the required time.

Timed Specimens

They are often used to monitor the level of a specific substance or condition in the patient. Blood is drawn at specific times for different reasons. They are:

- To measure blood levels of substances exhibiting diurnal variation. (e.g., cortisol hormone)
- To determine blood levels of medications (e.g., digoxin for cardiovascular disease)
- To monitor changes in a patient's condition (e.g., steady decrease in hemoglobin level)

Two-Hour Postprandial Test

This test is used to evaluate diabetes mellitus. Fasting glucose level is compared with the level 2 hours after eating a full meal or ingesting a measured amount of glucose.

Oral Glucose Tolerance Test (OGTT)

This test is used to diagnose diabetes mellitus and evaluate patients with frequent low blood sugar. 3-hour OGTT is used to test hyperglycemia (abnormally high blood sugar level) and diagnose diabetes mellitus. 5-hour OGTT is used to evaluate hypoglycemia (abnormally low blood sugar level) for disorders of carbohydrate metabolism.

OGTT are scheduled to begin between 0700 and 0900.

Blood Cultures (BC)

They are ordered to detect presence of microorganisms in the patient's blood. The patient will usually have chills and fever of unknown origin (FUO), indicating the possible presence of pathogenic microorganisms in the blood (septicemia).

Blood cultures are usually ordered STAT or as timed specimen, and collection requires strict aseptic technique.

PKU

Test ordered for infants to detect phenylketonuria, a genetic disease that causes mental retardation and brain damage. Test is done on blood from newborn's heel or on urine.

Special Specimen Handling

Cold Agglutinins

Cold agglutinins are antibodies produced in response to Mycoplasma pneumoniae infection (atypical pneumonia). The antibodies formed may attach to red blood cells at temperatures below body temperature, and as such, the specimen must be kept warm until the serum is separated from the cells.

Blood is collected in red-topped tubes pre-warmed in the incubator at 37 degrees Celsius for 30 minutes.

Chilled specimens

Some tests require that the specimen collected be chilled immediately after collection in crushed ice or ice and water mixture. Likewise, the specimen must be immediately transported to the laboratory for processing.

Some of the tests that require chilled specimen are: arterial blood gases, ammonia, lactic acid, pyruvate, ACTH, gastrin, and parathyroid hormone.

Light-sensitive specimens

Specimens are protected from light by wrapping the tubes in aluminum foil immediately after they are drawn. Exposure to light could alter the test results for: Bilirubin, beta-carotene, Vitamins A & B6, and porphyrins.

Dermal Punctures (Microcapillary collection)

When venipuncture is inadvisable, it is possible to perform a majority of laboratory tests on micro samples obtained by dermal (skin) puncture, with the exception of ESR, coagulation studies, blood cultures and other tests that require a large amount of serum. Dermal puncture may be done on both pediatric and adult patients.

Punctures should never be performed with a surgical blade or hypodermic needle because they can be difficult to control. Deep penetration into the skin can cause serious injury such as osteomyelitis (inflammation of the bone and bone marrow). A lancet should be used, which delivers a pre-determined depth that can range from 0.85mm for infants to 3.0mm for adults.

Site selection for infant microcapillary collection

The heel is used for dermal punctures on infants less than 1 year of age. Areas recommended are the medial and lateral areas of the plantar surface of the foot. These are determined by drawing imaginary lines medially extending from the middle of the great toe to the heel and laterally from the middle of the fourth and fifth toes to the heel.

The depth may vary according to different textbooks; however, the NHA follows the American Academy of Pediatrics recommendation that heel punctures for infants not exceed 2.0mm.

ORDER OF DRAW

Often requests are for more than one test to be performed; and as such, more than one collection tube needs to be drawn. According to the National Committee for Clinical Laboratory Standards, (NCCLS, 2003 Guidelines) the correct order of draw is:

1. Blood Cultures
2. Light Blue top tubes
3. Serum or non-additive tube (Red or Red/Gray top tubes)
4. Green top tubes
5. Lavender top tubes
6. Gray top tubes

To help their students memorize the new order of draw, the staff at Phlebotomy Education LLC in Allen Park MI have put together this simple sentence mnemonic: "BeCause Better Specimens Generate Perfect Goals."

BeCause	=	Blood Cultures
Better	=	Blue
Specimens	=	Serum (Red)
Generate	=	Green
Perfect	=	Purple (Lavender)
Goals	=	Gray

Reprinted with Permission from: Phlebotomy Education, LLC

Nancy Glasgow, CPT/RHE (NHA); Brandie L.B. Jonson, CPT/RHE (NHA); Joan Yarbrough, CPT/RHE (NHA)

As you know, it is important for you, the Healthcare Professional to stay current with changes in your industry. Therefore, the National Healthcareer Association and the certification exam, Certified Phlebotomy Technician, CPT, will only accept this as the correct order of draw. It is important for you, the student, to know that since these guidelines are relatively new, your textbook may not have presented this as the correct order of draw.

Note: 1998 National Committee for Clinical Laboratory Standards, (NCCLS) guidelines prescribe the same order for filling syringe samples as direct filling from a multi sample needle.

TEST TUBES, ADDITIVES AND TESTS

⊘ *Lavender top tube -*

Contains the anticoagulant ethylenediaminetetraacetic acid (EDTA).
EDTA inhibits coagulation by binding to calcium present in the specimen.

The tubes must be filled at least two-thirds full and inverted eight times.

Common tests: CBC (Complete Blood Count)
Includes: RBC count, WBC count and Platelet count; WBC
differential count
Hemoglobin and Hematocrit determinations
ESR (Erythrocyte Sedimentation Rate)
Sickle Cell Screening

⊘ *Light-Blue top tube -*

Contains the anticoagulant Sodium Citrate, which also prevents
coagulation by binding to calcium in the specimen. Sodium citrate is the
anticoagulant used for coagulation studies because it preserves the
coagulation factors.

The tube must be filled completely to maintain the ratio of nine parts
blood to one part sodium citrate, and should be inverted three to four
times.

Common tests: Coagulation Studies-
Prothrombin Time (PT) – evaluates the extrinsic
system of the coagulation cascade and monitors
Coumadin therapy.
Activated Partial Thromboplastin Time (APTT,
PTT) - Evaluates the intrinsic system of the
coagulation cascade and monitors Heparin therapy.
Fibrinogen Degradation Products (FDP)
Thrombin Time (TT)
Factor assays
Bleeding Time (BT)

⊘ Green top tube -

Contains the anticoagulant Heparin combined with sodium, lithium, or ammonium ion. Heparin works by inhibiting thrombin in the coagulation cascade. It is not used for hematology because heparin interferes with the Wright's stained blood smear.

This tube should be inverted eight times.

> **Common tests**: Chemistry tests: performed on plasma such as Ammonia, carboxyhemoglobin & STAT electrolytes.

⊘ Gray top tube –

Contains additives and anticoagulants. All gray top tubes contain glucose preservative (antiglycolytic agent): sodium fluoride- preserves glucose for 3days; or lithium iodoacetate- preserves glucose for 24 hours. May also contain the anticoagulant potassium oxalate, which prevents clotting by binding calcium.

This tube should be inverted eight times.

> **Common tests:** Fasting blood sugar (FBS)
> Glucose tolerance test (GTT)
> Blood alcohol levels
> Lactic acid measurement

⊘ Red/Gray (speckled) top tube -

Also called tiger-top tube and serum separator tubes (SST)

Contain clot activators: glass particles, silica and celite which hastens clot formation, and thixotropic gel, a serum separator which when centrifuged forms a barrier between the serum and the cells preventing contamination of the serum with cellular elements.

Tubes must be inverted five times.

> **Common tests:** Most chemistry tests

◊ *Red top tube -*

Also known as plain vacuum tube and contains no additive or anticoagulant. Collected blood clots by normal coagulation process in 30 minutes.

There is no need to invert the tube after collection.

Common tests – Serum chemistry tests
Serology tests
Blood bank

◊ **Yellow top tube -** (sterile)

Contains the anticoagulant sodium polyanetholesulfonate (SPS). These are used to collect specimens to be cultured for the presence of microorganisms. The SPS aids in the recovery of microorganisms by inhibiting the actions of complement, phagocytes, and certain antibiotics.

These tubes should be inverted eight times.

Hemostasis

Hemostasis is the process by which blood vessels are repaired after injury. This is a process that starts from vascular contraction as an initial reaction to injury, then to clot formation, and finally removal of the clot when the repair to injury is done. It occurs in four stages:

Stage 1: Vascular phase
Injury to a blood vessel causes it to constrict slowing the flow of blood.

Stage 2 – Platelet phase
Injury to the endothelial lining causes platelets to adhere to it. Additional platelets stick to the site finally forming a temporary platelet plug in a process called 'aggregation'.

Vascular phase and platelet phase comprise the primary hemostasis. Bleeding time test is used to evaluate primary hemostasis.

Stage 3 – Coagulation phase

This involves a cascade of interactions of coagulation factors that converts the temporary platelet plug to a stable fibrin clot. The coagulation cascade involves an intrinsic system and extrinsic system, which ultimately come together in a common pathway.

Activated partial thromboplastin time (APTT) – test used to evaluate the intrinsic pathway. This is also used to monitor heparin therapy.

Prothrombin time (PT) – test used to evaluate the extrinsic pathway. This is also used to monitor coumadin therapy.

Stage 4 – Fibrinolysis

This is the breakdown and removal of the clot. As tissue repair starts, plasmin (an enzyme) starts breaking down the fibrin in clot. Fibrin degradation products (FDPs) measurement is used to monitor the rate of fibrinolysis.

Needle Stick Prevention Act

OSHA has put into force the Occupational Exposure to Bloodborne pathogen (BBP) Standard when it was concluded that healthcare employees face a serious health risk as a result of occupational exposure to blood and other body fluids and tissues. The standards outline necessary engineering and work practice controls that OSHA believes will help minimize or eliminate exposure to employees. The standard was revised in 2001 to conform to the Needlestick Safety and Prevention Act passed in November 2000. The act directed OSHA to revise the BBP standard in four key areas:

- Revision and updating of the exposure control plan.
- Solicitation of employee input in selecting engineering and work practice controls.
- Modification of definitions relating to engineering controls (i.e., sharps disposal containers, self-sheathing needles, needleless systems.
- New record keeping requirements.

The employer must establish and maintain a sharps injury log for percutaneous injury from contaminated sharps and it must be done in such a manner to protect the confidentiality of the injured employee.

The sharps injury log must contain, at a minimum:
   a. The type and brand of device involved in the incident.
   b. The department or work area where the exposure incident occurred.
   c. An explanation of how the incident occurred.

** See Appendix B for further clarification.

Latex Sensitivity

Latex sensitivity is an emerging and important problem in the health care field. Following the development of Universal Precaution Standards (OSHA, 1980), the use of natural rubber latex gloves for infection control skyrocketed. Within the last decade, however, the incidence of latex sensitivity has grown. It is an issue that every health care worker must be concerned about. Individuals with a known sensitivity to latex should wear a medical alert bracelet.

Introduction to Microbiology

Safety Considerations

Infectious and contagious diseases can be transmitted easily from person to person by direct or indirect contact. For this reason, it is important that health care personnel strictly adhere to the OSHA standards pertaining to bio-hazardous materials. Health care workers, both inside and outside the laboratory, can become exposed to agents capable of causing infectious disease on a daily basis. Agents, such as bacteria, fungi, parasites, and viruses, are capable of causing serious disease; so strict adherence to safety rules is required to prevent disease transmission.

Microbiology laboratories contain appropriate safety equipment. Biological safety cabinets and biohazard devices that encase a work area to protect laboratory personnel from accidental exposure to infectious diseases. Air that contains infectious materials can be decontaminated by passing it through a high-efficiency particulate air (HEPA) filter or by exposure to ultraviolet light or heat. In addition to gloves, laboratory coats and other protective clothing should be worn over conventional street apparel. Moreover, certain activities in the laboratory may require additional protective clothing. Strict decontamination procedures must always be followed.

All materials containing potentially infectious agents must be decontaminated and/or disposed of according to OSHA standards. Biohazard containers should be used. Workbenches and other surface areas are cleaned with liquid antiseptic agents. A 10% solution of sodium hypochlorite (household bleach in water) is an effective agent against viral contamination. If a prepared antiseptic solution is used, a new mixture should be prepared every 7 days to maintain effectiveness. When using commercial antiseptics, always check the expiration date. Reusable materials must be sterilized in an autoclave according to the manufacturer's specifications.

Smear Preparation, Staining Techniques, and Wet Mounts

The Gram Stain

The Gram Stain is used to classify bacteria on the basis of their form, size, cellular morphology, and Gram Stain reaction. It is a critical test for the rapid presumptive identification of infectious agents, and it also is a means by which the quality of clinical specimen can be evaluated. Hans Christian Gram originally developed the test in 1884. In 1921, Hucker modified it to the test we use today.

When exposed to the Gram Stain, bacteria stain either gram-positive (deep violet) or gram-negative (light to dark red) on the basis of differences in cell wall composition and structure. Gram-positive bacteria have a thick peptidoglycan layer and large amounts of teichoic acids. This combination prevents them from being affected by alcohol decolorization; therefore, they retain the initial stain of crystal violet, which imparts a deep violet color. Gram-negative cell walls have a single peptidoglycan layer attached to a symmetric, lipoplysaccharide, phospholipid, bilayered, outer membrane interspersed with protein. The outer membrane is damaged by the alcohol decolorizer, allowing the crystal violet iodine complex to leak out and be replaced by the Sefranin counterstain (red). The Gram stain can be affected by many factors, including culturing, age, antibiotics, the medium in which the bacteria is growing, incubation, atmosphere, phagocytosis, and staining technique.

Smear Preparation

Proper smear preparation will produce a thin monolayer of organisms for easy visualization but will be thick enough to reveal characteristic arrangements of the bacteria. Always wear latex gloves and a laboratory coat and follow all other universal precautions when handling clinical specimens.

Pre-cleaned, glass slides with frosted ends should be used for the smear. The frosted ends are desirable as they allow accurate labeling and convenient handling. Frequently, a direct smear is prepared from the swab used to obtain the sample. A smear can be from any body opening, including the genitals or wounds (such as surgical sites, bites, cuts, or body ulcers). The best process is to obtain two swabs, one for the culture and one for the smear. If this is the case, the specimen is cultured first. Then, before the thioglycolate tube is inoculated, the smear is prepared. The danger in using one swab is that the target area may be missed, thus invalidating the entire testing process. You will want to check laboratory protocols for smear preparation to determine the exact procedure for obtaining a smear specimen.

Smearing and Fixation Technique

To prepare the smear, gently roll the swab across the slide, in one direction, leaving a thin film of specimen material on the slide. Specimens not received on swabs can be spread over a large area by using sterile swabs or a heat-sterilized wire loop to form a thin film on the slide. Extremely thick specimens can be placed on one slide, covered with a second slide, and pulled apart. The excess on the edge of the slide can be removed using a disinfectant-soaked paper towel. The smearing and fixation technique must be done in a bio safety cabinet.

A more commonly used technique for thick specimens is to place a drop of saline on the slide to facilitate smear preparation. Smears of cerebro spinal fluid or other body fluids requiring centrifugation may be prepared by using a Cytospin slide centrifuge to concentrate the fluids. This method is used to increase the likelihood of visualizing bacteria and to decrease examination time for more rapid results. Using slides with etched rings helps to locate the inoculated area.

Smears should be air-dried on a flat surface or on an electric slide warmer heated to 60 degrees Centigrade. The slide is placed on the supporting rods of the stain rack and then fixed by covering the slide with methanol for 1 minute. The residual methanol is then drained off without rinsing and is allowed to air-dry again. The slide is then ready to stain. Do no heat-fix the slide before staining. Methanol fixation is preferred over the old standard of heat-fixing smears because it prevents lysis of red blood cells (RBCs), gives a cleaner background, does not affect bacterial morphology, and is safer.

Staining Bacteria

Gram staining is routinely performed in the microbiology section of the clinical laboratory. Bacteria (microorganisms) are so tiny and virtually colorless that their morphology cannot be determined without first staining them. The Gram stain is the universal technique used to stain bacteria after they have been methanol-fixed.

The staining procedure involves the sequential application of primary stain mordant, decolorizer, and counterstain to a bacterial smear. The organisms according to the chemical composition of the cell walls take up the stains differently. A fixed smear is placed on a staining rack and the primary stain crystal violet is poured onto one end of the smear until the whole side is covered. The stain is allowed to remain in one place for 30 seconds.

Staining of Blood Smears

The stain commonly used for examination of blood cells is called polychromatic because they contain dyes that will stain various cell components different colors. These stains usually contain methylene blue, a blue stain, and eosin, a re-orange stain. These stains are attracted to different parts of the cell. Thus, the cells and their structures can be more easily visualized and differentiated. The most commonly used differential bloodstain is wright's stain.

Wright's stain is applied to the slide for approximately 1 to 3 minutes. Butter is added on top of the stain and is mixed by gently blowing until a green metallic sheen appears; this usually takes 2 to 4 minutes. The slide is then gently rinsed and is allowed to air-dry. A properly stained smear should appear pinking to the naked eye.

Semi-automated slide stainers are frequently used in large laboratories. These machines are capable of staining a large number of blood smears with consistency and reliability. Small laboratories generally use a manual quick stain method. With this three-step method, the smears are exposed to Wright's stains, butter, and distilled water in a timed sequenced.

Manual staining techniques. The slide is immersed first in Wright's stain, then in butter solution. It is then rinsed in distilled water and placed in a drying rack.

Urinalysis

Urine Formation

The kidney is a highly discriminating organ that maintains the internal environment by selectively excreting or retaining various substances in response to specific body needs. Blood enters the glomerulus of each nephron through the afferent arteriole and flows into the glomerular capillaries.

The walls of the glomerulus capillaries are highly permeable to water and various dissolved substances from blood plasma. These substances filter through the capillary walls and pass into the tubule, where reabosorbtion of some substances, secretion of others, and the concentration of others occur.

Certain components, such as glucose, water, and amino acids, are partially or completely reabsorbed by the capillaries surrounding the proximal tubules. In the distal tubule, additional water is absorbed, and potassium and hydrogen are secreted. The urine is concentrated in the loop of Henle and the collecting tubules. The average daily output of urine ranges from 1000 to 2500 mL, depending on the individual's state of hydration. The average adult feels the need to urinate when the bladder contains approximately 300 to 400 mL of urine. Tissue hydration correlates directly to fluid intake and output and fluid loss and retention.

Red Urine

The most frequently observed color abnormally is red or red-brown. When the urine appears to be a cloudy pink, red, or red-brown, it may indicate hematuria, which is the presence of red blood cells in the urine. Hemoglobinuria, the presence of hemoglobin in the urine, results in a clear pink, red, or red-brown color. The presence of porphyrins, a group of pigments associated with the production of hemoglobin, is usually responsible for a red or purple color. The presence of blood in urine with an acid pH may result in a dark brown or even black color. The presence of red blood cells or hemoglobin in the urine is easily confirmed during chemical and microscopic examination.

Urinalysis (UA) is a laboratory procedure that has two purposes, one of which is to detect body disturbances, such as endocrine or metabolite abnormalities. Its second purpose is to detect intrinsic conditions that may adversely affect the kidneys and urinary tract. The composition of urine gives the physician valuable clues as to the well being of the urinary system and other body systems. A routine UA is an examination of urine to determine the presence of abnormal elements, which may indicate various pathologic conditions.

Basic Procedural Steps
- Inspect physical characteristics.
- Analyze urine components through chemical measurements.
- Perform a microscopic examination.

Collecting the Urine Specimen

Whether a specimen is a secretion, excretion, or body material obtained at biopsy, the validity of the results depends on proper specimen collection. Examination of the urine specimen can yield results that will assist in the diagnosis and treatment of the patient. Although medical assistants often collect a variety of specimens for testing outside the physician's office, this chapter focuses on the urine tests that are usually performed within the medical office/clinic.

The manner in which the specimen is collected depends on the test to be performed. The health care worker must adhere to proper urine collection techniques and must be certain to obtain the proper specimen as ordered by the physician. Some specimen collections are complicated and require specifically trained personnel and extensive patient preparation. In the medical office, the collection, processing, and/or transport of most specimens can be accomplished without complication. Patient education is the responsibility of the medical assistant, and collecting a urine specimen requires clear and concise instructions.

General Instructions for Urine Collection

Urine specimens may be collected in the medical office or at home. In either situation, it is important to follow appropriate procedures for specimen collection and processing.

Instructions for Urine Collection
- o Carefully label all specimens. Do not apply the label to the container lid, but place the label on the container itself. Use an indelible marker or make sure that the label will adhere to the container at refrigerated temperatures. On the label, record the patient's name, the date and time of collection, and the type of specimen. Add the physician's name if the specimen is to be sent to a central laboratory facility opening that is 2 inches in diameter. If the specimen is to be obtained from a pediatric patient, the container may be slightly smaller. If the specimen is to be transported, be sure it has a screw-type lid.
- o If a bacterial culture is ordered, make sure a sterile container is available. If this is the case, the specimen may have to be obtained through catheterization.
- o Advise female patients, with the consent of the physician, that the collection of a urine specimen should be avoided, if possible, during their menstrual cycle and for several days before and after, as the specimen may be contaminated with blood.
- o If the analyte is unstable or if the testing is delayed, you may add preservatives to the specimen. Check your laboratory's procedure manual or the procedural manual provided by the referral laboratory to determine the proper preservative for each test. Remember that the preservative must not interfere with the test procedure or results. Always not on the specimen the type and amount of preservative added.

Types of Specimen Collection

First Morning Sample

A first morning sample is the type of specimen most commonly used for routine urinalysis. Because the concentration of urine varies throughout the day, it is usually easiest to identify abnormalities in a relatively concentrated specimen. The first morning specimen may also be called an early morning specimen, as it represents the urine formed over approximately an 8-hour period.

Because it is impractical to collect a first morning specimen in the medical office, the patient must be instructed in the proper collection technique for a clean-catch or mid-stream urine sample. The specimen can then be collected at home and brought to the office. Be certain that the patient knows to refrigerate the specimen until it is transported to the office. (Alternatively, the patient may be instructed to add a preservative to the container.) The laboratory should supply the container, and any preservatives to be added, as a container from home may not be properly washed and rinsed prior to use.

When the specimen is delivered to the office or laboratory, the medical assistant should check it for proper labeling and perform the required test(s) immediately. If that is not possible, the specimen may be refrigerated until testing can be done.

Mid-Stream Specimen

A mid-stream urine specimen is one that is collected not at the beginning or end of voiding, but in the middle of urination. The patient is instructed to void the first one third of the urine into the toilet. At the point, the patient stops urine flow, places the specimen container into position, and voids the next one third of the urine into the container. Once the specimen is collected, the patient can then finish emptying the bladder into the toilet. The specimen volume should be at least 25 mL of urine. A mid-stream specimen is thought to be a better representative of the contents of the bladder because it is free from the contaminants that may have been in the urethra or the urinary meatus.

Clean-Catch Specimen

Most laboratories prefer a clean-catch, mid-stream specimen for testing, as it provides the clearest, most accurate results. If the urine specimen is to be tested for bacteria or antibiotic sensitivity and a catheterized specimen is not required, a clean-catch sample will be needed. Collecting this sample requires special cleaning of the external genitalia. Because most patients are not familiar with aseptic technique, they must be carefully instructed on the procedure. In the case of a disabled or elderly individual, assistance may be needed in obtaining the specimen.

24-Hour Urine Collection (Addis Test)

Because the concentration of urine changes over a 24-hour period, a single urine sample may not yield results that are representative of the patient's true clinical picture. The Addis test requires pooling of urine over a 3-, 6-, 12-, or 24-hour period. The method of collection is the same for all time periods; only the length of time that the sample is collected varies. The primary purpose of this test is to determine the quantity of an analyte in the specimen. The normal volume of urine produced every 24 hours varies. Infants and children produce smaller volumes than adults.

Because the urine will not be tested until the completion of the designated collection time, it is necessary to add a preservative and to refrigerate the urine to avoid decomposition. Consult the procedural manual for the proper preservative, as it must be one that does not interfere with the analyte.

Time is an important parameter in the Addis test. Most patients find it easiest to begin this test upon arising; however, the test may be begun at any time. The first morning void is not collected for testing.

Normal 24-Hour Urine Volumes	
AGE OF PATIENT	URINE VOLUME (mL/24hr)
Neonate	20-350
Child (1-9 yrs old)	300-600
Adolescent (10-16 yrs old)	600-1500
Adult	600-2000

Specific Gravity

The specific gravity of urine is the ratio of the weight of a given volume of urine to the weight of the same volume of distilled water at a constant temperature. Specific gravity is the most convenient way of measuring the kidneys' ability to concentrate and dilute. An abnormality in the ability of the kidney to concentrate or dilute urine is an indication of renal disease or hormonal deficiency.

During a 24-hour period, normal adults with normal diets and normal fluid intake produce urine with a specific gravity of between 1.015 and 1.025. The normal range of urine specific gravity for a random collection is 1.005 to 1.030.

Urine Volume

The normal volume of urine produced by an adult in a 24-hour period is 600 to 2000 mL. Polyuria is a 24-hour urine output that exceeds 2000 mL. In normal patients, the predominant factor affecting urine output is water intake. Urinary output is measured by doing a 24-hour urine collection.

Urinary pH

The pH, or the percentage of hydrogen ion concentration of a solution, is a reflection of the acidity or alkaline of a solution. A pH of 7.0 is considered to be neutral. The pH of distilled water is 7.0. A pH of 0 to 7.0 is considered to be acidic, whereas a pH of 7 to 14 is considered to be alkaline or basic.

Normal, freshly voided urine will usually have a pH of 4.5 to 8.0. Within this range, the urine pH of most healthy patients is around 6.0.

An accurate measurement of urinary pH requires a freshly voided urine specimen. Unless refrigerated, urine becomes alkaline upon standing owing to the loss of carbon dioxide and the conversion of urea into ammonia by bacteria. The pH portion of the reagent strip uses the indicators methyl red and bromthymol blue. These indicators show a color range that changes from orange to green to blue as the pH becomes more basic.

Urinary Glucose

Glucose is the sugar typically found in urine. Other sugars, such as lactose, fructose, galactose, and pentose, may be detected in urine under specific circumstances. Glucose is present in urine when the blood glucose level exceeds the renal threshold. Glycosuria is the presence of glucose in the urine.

Patients with diabetes mellitus have glycosuria, along with polynuria and thirst. The reagent strip test for glucose relies on enzymatic tests that are specific for glucose. A common reagent strip urinary glucose enzymatic method uses glucose oxidase. The glucose oxidase reacts specifically with glucose. Sugars, such as lactose, fructose, and others, are not detected by the glucose oxidase method. A copper reaction test is a commonly used confirmatory and screening test for glucose and other reducing substances in urine. Copper reduction tests are used in pediatric patients to detect increased levels of glucose that may not be detected by the specific enzymatic test found on most reagent strips.

Urinary Bacteria

Enteric gram-negative bacteria that are always nitrite positive can convert urinary nitrate to nitrite. A positive nitrite test is an indication that a significant number of bacteria are present in the urine. Bacteriuria is the presence of bacteria in the urine.

Urinary Leukocytes

The presence of increased numbers of leukocytes or white blood cells in the urine is an indicator of bacteriuria or urinary tract infection (UTI). Granulocytic leukocytes release esterase when the cells lyse. Testing for leukocyte esterase by the reagent strip method is used in tandem with the microscopic examination of urine sediment for the diagnosis of bacteriuria or UTI.

A positive test by the reagent strip method is indicated by a purple color. The greater the amount of leukocytes/esterase present, the greater the intensity of the purple color. Bacterial culture and sensitivity testing best confirm UTI's. A clean-catch mid-stream urine sample is usually required for any bacterial culture. For this reason, it is ALWAYS wise to collect a clean-catch urine specimen; do not dispose of the specimen until the physician directs you to do so.

Specialized Urine Tests/Urinary Pregnancy Testing

Probably the most common specialized urine test is the pregnancy test. Human chorionic gonadotropin (hCG), also known as uterine chorionic gonadotropin (UCG), is produced in the placenta and is detectable in the blood and urine early in the gestation period. HCG is not normally found in the urine of young, healthy, non-pregnant women. Because of hCG's early appearance during gestation, increased levels of hCG are a natural marker for pregnancy.

APPENDIX A: PATIENTS BILL OF RIGHTS

As a patient in XXX Hospital you have the right, consistent with law, to:

1. Receive treatment without discrimination as to race, color, religion, gender, national origin, disability, or source of payment.
2. Receive considerate and respectful care in a clean and safe environment free of unnecessary restraints.
3. Receive emergency care if you need it.
4. Be informed of the name and position of the doctor who will be in charge of your care in the hospital.
5. Know the names, positions and functions of any hospital staff involved in your care.
6. Receive complete information about your diagnosis, treatment and prognosis.
7. Receive all the information that you need to give informed consent for any proposed procedure or treatment. This information shall include the possible risks and benefits of the procedure or treatment.
8. Receive all the information you need to give informed consent for an order not to resuscitate. You also have the right to designate an individual to give this consent for you if you are too ill to do so. If you would like additional information, please ask
9. Refuse treatment, examination, or observation, if retired or a family member, and be told what effect this may have on your health.
10. Refuse to take part in research. In deciding whether or not to participate, you have the right to a full explanation.
11. Privacy while in the hospital and confidentiality of all information and records regarding your care.
12. Participate in all decisions about your treatment and discharge from the hospital.
13. Review your medical record without charge. Obtain a copy of your medical record for which the hospital can charge a reasonable fee. You cannot be denied a copy solely because you cannot afford to pay.
14. Receive a bill and explanation of all charges.
15. Complain without fears of reprisals about the care and services you are receiving and to have the hospital respond to you; and if requested, a written response. If you are not satisfied with the hospital's response, you can complain to the Patient Representative Office located here in the hospital.
16. Receive information about pain and pain relief measures, be involved in pain management plan, and receive a quick response to reports of pain.
17. Receive healthcare in an environment that is dedicated to avoiding patient harm and improving patient safety.
18. The right to request information about advance directives regarding your decisions about medical care.
19. Make known your wishes in regard to anatomical gifts. Your may document your wishes in your health care proxy or on a donor card, available from the hospital.
20. Understand and use these rights. If for any reason you do not understand or you need help, the hospital will attempt to provide assistance, including an interpreter.

Patient Responsibilities

Provision of Information: You have the responsibility to provide, to the best of your knowledge, accurate and complete information about present complaints, past illness, hospitalizations, medications, and other matters relating to your health. You have the responsibility to report unexpected changes in your condition to the responsible practitioner. You are responsible for making it known whether you clearly comprehend a contemplated course of action and what is expected of you.

Compliance with Instructions: You are responsible for following the treatment plan recommended by the practitioner primarily responsible for your care. This may include following the instructions of nurses and allied health personnel as they carry out the coordinated plan of care and implement the responsible practitioner's orders, and as they enforce the applicable hospital rules and regulations. You are responsible for keeping appointments and, when you are unable to do so for any reason, for notifying the responsible practitioner or the hospital.

Refusal of Treatment: You are responsible for your actions if you refuse treatment or do not follow the practitioner's instructions.

Hospital Rules and Regulations: You are responsible for following hospital rules and regulation affecting patient care and conduct.

Respect and Consideration: You are responsible for being considerate of the rights of other patients and hospital personnel and for assisting in the control of noise, smoking and the number of visitors. You are responsible for being respectful of the property of other persons and the hospital.

Patient Representative

The Patient Representative's primary assignment is to assist you in exercising your rights as a patient. He/she is also available to act as your advocate and to provide a specific channel through which you can seek solutions to problems, concerns and unmet needs. You may call the Patient Representative at (000)000-0000.

The Patient Bill of Rights for Pain Management

You have the right to:

- Information about pain and pain relief
- A caring staff who believe your reports of pain
- A care staff with concern about your pain
- A quick response when you report your pain

You have the responsibility to:

- Ask for pain relief when your pain first starts
- Help those caring for you to assess your pain
- Tell those caring for you if your pain is not relieved
- Tell those caring for you about any worries that you have about taking pain medications
- Decide if you want your family and/or significant others to aid in your relief of pain

APPENDIX B: OSHA REGULATIONS

OSHA Occupational Safety & Health Administration
U.S. Department of Labor

Home
Index
Search

Revision to OSHA's Bloodborne Pathogens Standard
Technical Background and Summary

April 2001

Background

The Occupational Safety and Health Administration published the Occupational Exposure to Bloodborne Pathogens standard in 1991 because of a significant health risk associated with exposure to viruses and other microorganisms that cause bloodborne diseases. Of primary concern are the human immunodeficiency virus (HIV) and the hepatitis B and hepatitis C viruses.

The standard sets forth requirements for employers with workers exposed to blood or other potentially infectious materials. In order to reduce or eliminate the hazards of occupational exposure, an employer must implement an exposure control plan for the worksite with details on employee protection measures. The plan must also describe how an employer will use a combination of engineering and work practice controls, ensure the use of personal protective clothing and equipment, provide training, medical surveillance, hepatitis B vaccinations, and signs and labels, among other provisions. Engineering controls are the primary means of eliminating or minimizing employee exposure and include the use of safer medical devices, such as needleless devices, shielded needle devices, and plastic capillary tubes.

Nearly 10 years have passed since the bloodborne pathogens standard was published. Since then, many different medical devices have been developed to reduce the risk of needlesticks and other sharps injuries. These devices replace sharps with non-needle devices or incorporate safety features designed to reduce injury. Despite these advances in technology, needlesticks and other sharps injuries continue to be of concern due to the high frequency of their occurrence and the severity of the health effects.

The Centers for Disease Control and Prevention estimate that healthcare workers sustain nearly 600,000 percutaneous injuries annually involving contaminated sharps. In response to both the continued concern over such exposures and the technological developments which can increase employee protection, Congress passed the **Needlestick Safety and Prevention Act** directing OSHA to revise the bloodborne pathogens standard to establish in greater detail requirements that employers identify and make use of effective and safer medical devices. That revision was published on Jan. 18, 2001, and became effective April 18, 2001.

Summary

The revision to OSHA's bloodborne pathogens standard added new requirements for employers, including additions to the exposure control plan and keeping a sharps injury log. It does not impose new requirements for employers to protect workers from sharps injuries; the original standard already required employers to adopt engineering and work practice controls that would eliminate or minimize employee exposure from hazards associated with bloodborne pathogens.

The revision does, however, specify in greater detail the engineering controls, such as safer medical devices, which must be used to reduce or eliminate worker exposure.

Exposure Control Plan

The revision includes new requirements regarding the employer's Exposure Control Plan, including an annual review and update to reflect changes in technology that eliminate or reduce exposure to bloodborne pathogens. The employer must:

- Take into account innovations in medical procedure and technological developments that reduce the risk of exposure (e.g., newly available medical devices designed to reduce needlesticks); and

- Document consideration and use of appropriate, commercially available, and effective safer devices (e.g., describe the devices identified as candidates for use, the method(s) used to evaluate those devices, and justification for the eventual selection).

No one medical device is considered appropriate or effective for all circumstances. Employers must select devices that, based on reasonable judgment:

- Will not jeopardize patient or employee safety or be medically inadvisable; and
- Will make an exposure incident involving a contaminated sharp less likely to occur.

Employee Input

Employers must solicit input from non-managerial employees responsible for direct patient care regarding the identification, evaluation, and selection of effective engineering controls, including safer medical devices. Employees selected should represent the range of exposure situations encountered in the workplace, such as those in geriatric, pediatric, or nuclear medicine, and others involved in direct care of patients.

OSHA will check for compliance with this provision during inspections by questioning a representative number of employees to determine if and how their input was requested.

Documentation of employee input

Employers are required to document, in the Exposure Control Plan, how they received input from employees. This obligation can be met by:

- Listing the employees involved and describing the process by which input was requested; or
- Presenting other documentation, including references to the minutes of meetings, copies of documents used to request employee participation, or records of responses received from employees.

Record keeping

Employers who have employees who are occupationally exposed to blood or other potentially infectious materials, and who are required to maintain a log of occupational injuries and illnesses under existing record keeping rules, must also maintain a sharps injury log. That log will be maintained in a manner that protects the privacy of employees. At a minimum, the log will contain the following:

- The type and brand of device involved in the incident;
- Location of the incident (e.g., department or work area); and
- Description of the incident

The sharps injury log may include additional information as long as an employee's privacy is protected. The employer can determine the format of the log.

Modification of Definitions

The revision to the bloodborne pathogens standard includes modification of definitions relating to engineering controls. Two terms have been added to the standard, while the description of an existing term has been amended.

Engineering Controls

Engineering Controls include all control measures that isolate or remove a hazard from the workplace, such as sharps disposal containers and self-sheathing needles. The original bloodborne pathogens standard was not specific regarding the applicability of various engineering controls (other than the above examples) in the healthcare setting. The revision now specifies that "safer medical devices, such as sharps with engineered sharps injury protections and needleless systems" constitute an effective engineering control, and must be used where feasible.

Sharps with Engineered Sharps Injury Protections

This is a new term which includes non-needle sharps or needle devices containing built-in safety features that are used for collecting fluids or administering medications or other fluids, or other procedures involving the risk of sharps injury. This description covers a broad array of devices, including:

- Syringes with a sliding sheath that shields the attached needle after use;
- Needles that retract into a syringe after use;
- Shielded or retracting catheters

- Intravenous medication (IV) delivery systems that use a catheter port with a needle housed in a protective covering.

Needleless Systems

This is a new term defined as devices that provide an alternative to needles for various procedures to reduce the risk of injury involving contaminated sharps. Examples include:

- IV medication systems which administer medication or fluids through a catheter port using non-needle connections; and
- Jet injection systems that deliver liquid medication beneath the skin or through a muscle.

Type Reaction	Symptoms/Signs	Cause	Prevention / Management
Irritant Contact Dermatitis	Scaling, drying, cracking of skin	Direct skin irritation by gloves, powder, soaps/detergents, incomplete hand drying	Obtain medical diagnosis, avoid irritant product, consider use of cotton glove liners, consider alternative gloves/products
Allergic Contact Dermatitis (Type IV delayed hypersensitivity **or** allergic contact sensitivity)	Blistering, itching, crusting (similar to poison ivy reaction)	Accelerators (e.g., thiurams, carbamates, benzothiazoles) processing chemicals (e.g., biocides, antioxidants) Consider penetration of glove barrier by chemicals	Obtain medical diagnosis, identify chemical. Consider use of glove liners such as cotton Use alternative glove material without chemical

			Assure glove material is suitable for intended use (proper barrier)
NRL Allergy - IgE/histamine mediated (Type I immediate hypersensitivity) ---------------------- A) Localized contact urticaria which may be associated with or progress to: B) Generalized Reaction	---------------------- Hives in area of contact with NRL ---------------------- Include: generalized urticaria, rhinitis, wheezing, swelling of mouth, and shortness of breath. Can progress to anaphylactic shock	NRL proteins: direct contact with or breathing NRL proteins, including glove powder containing proteins, from powdered gloves or the environment	Obtain medical diagnosis, allergy consultation, substitute non-NRL gloves for affected worker and other non-NRL products Eliminate exposure to glove powder - use of reduced protein, powder free gloves for coworkers Clean NRL-containing powder from environment Consider NRL safe environment

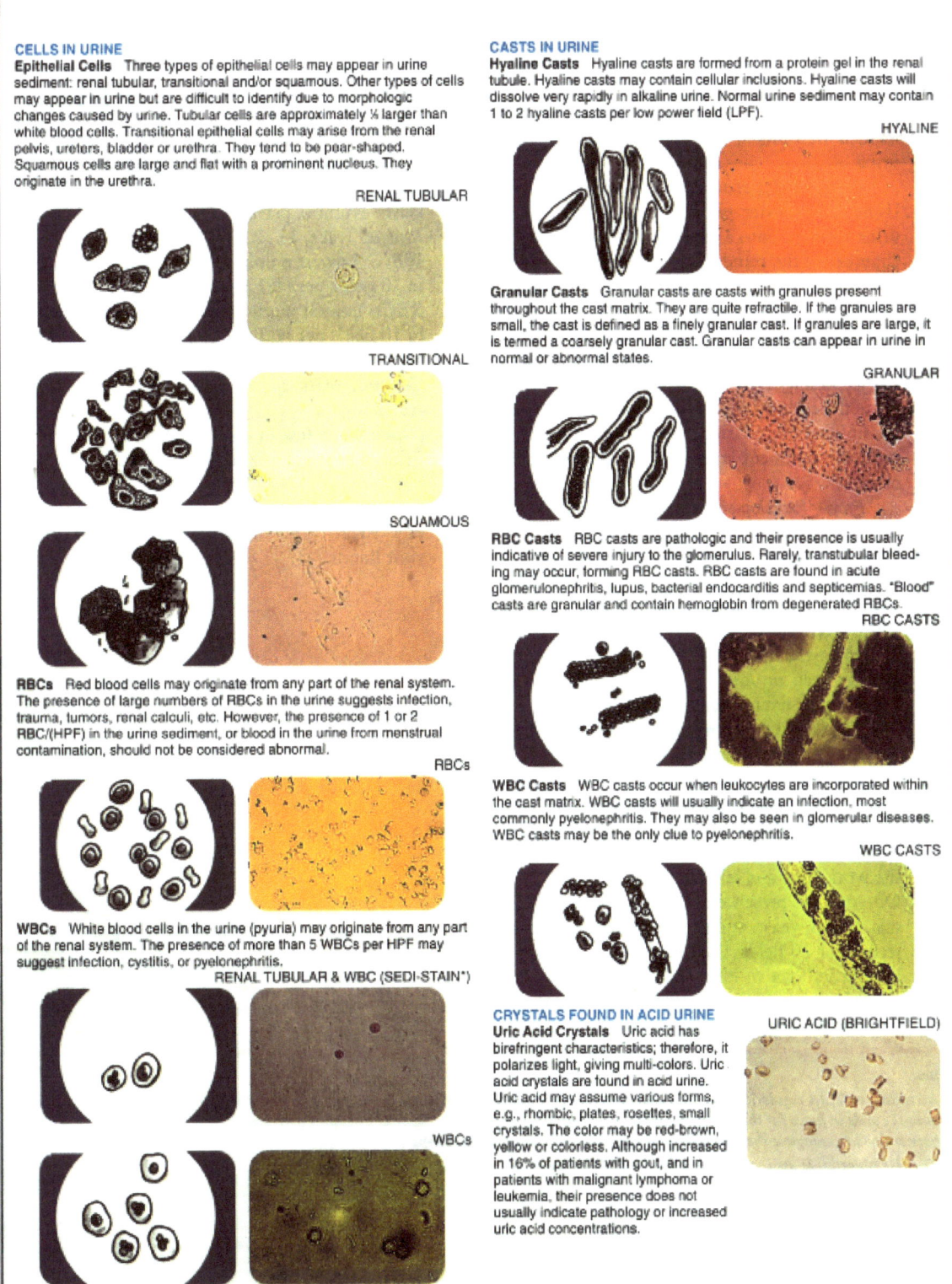

CELLS IN URINE

Epithelial Cells Three types of epithelial cells may appear in urine sediment: renal tubular, transitional and/or squamous. Other types of cells may appear in urine but are difficult to identify due to morphologic changes caused by urine. Tubular cells are approximately ⅓ larger than white blood cells. Transitional epithelial cells may arise from the renal pelvis, ureters, bladder or urethra. They tend to be pear-shaped. Squamous cells are large and flat with a prominent nucleus. They originate in the urethra.

RENAL TUBULAR

TRANSITIONAL

SQUAMOUS

RBCs Red blood cells may originate from any part of the renal system. The presence of large numbers of RBCs in the urine suggests infection, trauma, tumors, renal calculi, etc. However, the presence of 1 or 2 RBC/(HPF) in the urine sediment, or blood in the urine from menstrual contamination, should not be considered abnormal.

RBCs

WBCs White blood cells in the urine (pyuria) may originate from any part of the renal system. The presence of more than 5 WBCs per HPF may suggest infection, cystitis, or pyelonephritis.

RENAL TUBULAR & WBC (SEDI-STAIN*)

WBCs

CASTS IN URINE

Hyaline Casts Hyaline casts are formed from a protein gel in the renal tubule. Hyaline casts may contain cellular inclusions. Hyaline casts will dissolve very rapidly in alkaline urine. Normal urine sediment may contain 1 to 2 hyaline casts per low power field (LPF).

HYALINE

Granular Casts Granular casts are casts with granules present throughout the cast matrix. They are quite refractile. If the granules are small, the cast is defined as a finely granular cast. If granules are large, it is termed a coarsely granular cast. Granular casts can appear in urine in normal or abnormal states.

GRANULAR

RBC Casts RBC casts are pathologic and their presence is usually indicative of severe injury to the glomerulus. Rarely, transtubular bleeding may occur, forming RBC casts. RBC casts are found in acute glomerulonephritis, lupus, bacterial endocarditis and septicemias. "Blood" casts are granular and contain hemoglobin from degenerated RBCs.

RBC CASTS

WBC Casts WBC casts occur when leukocytes are incorporated within the cast matrix. WBC casts will usually indicate an infection, most commonly pyelonephritis. They may also be seen in glomerular diseases. WBC casts may be the only clue to pyelonephritis.

WBC CASTS

CRYSTALS FOUND IN ACID URINE

Uric Acid Crystals Uric acid has birefringent characteristics; therefore, it polarizes light, giving multi-colors. Uric acid crystals are found in acid urine. Uric acid may assume various forms, e.g., rhombic, plates, rosettes, small crystals. The color may be red-brown, yellow or colorless. Although increased in 16% of patients with gout, and in patients with malignant lymphoma or leukemia, their presence does not usually indicate pathology or increased uric acid concentrations.

URIC ACID (BRIGHTFIELD)

Figure 43-1. Atlas of urine sediment.

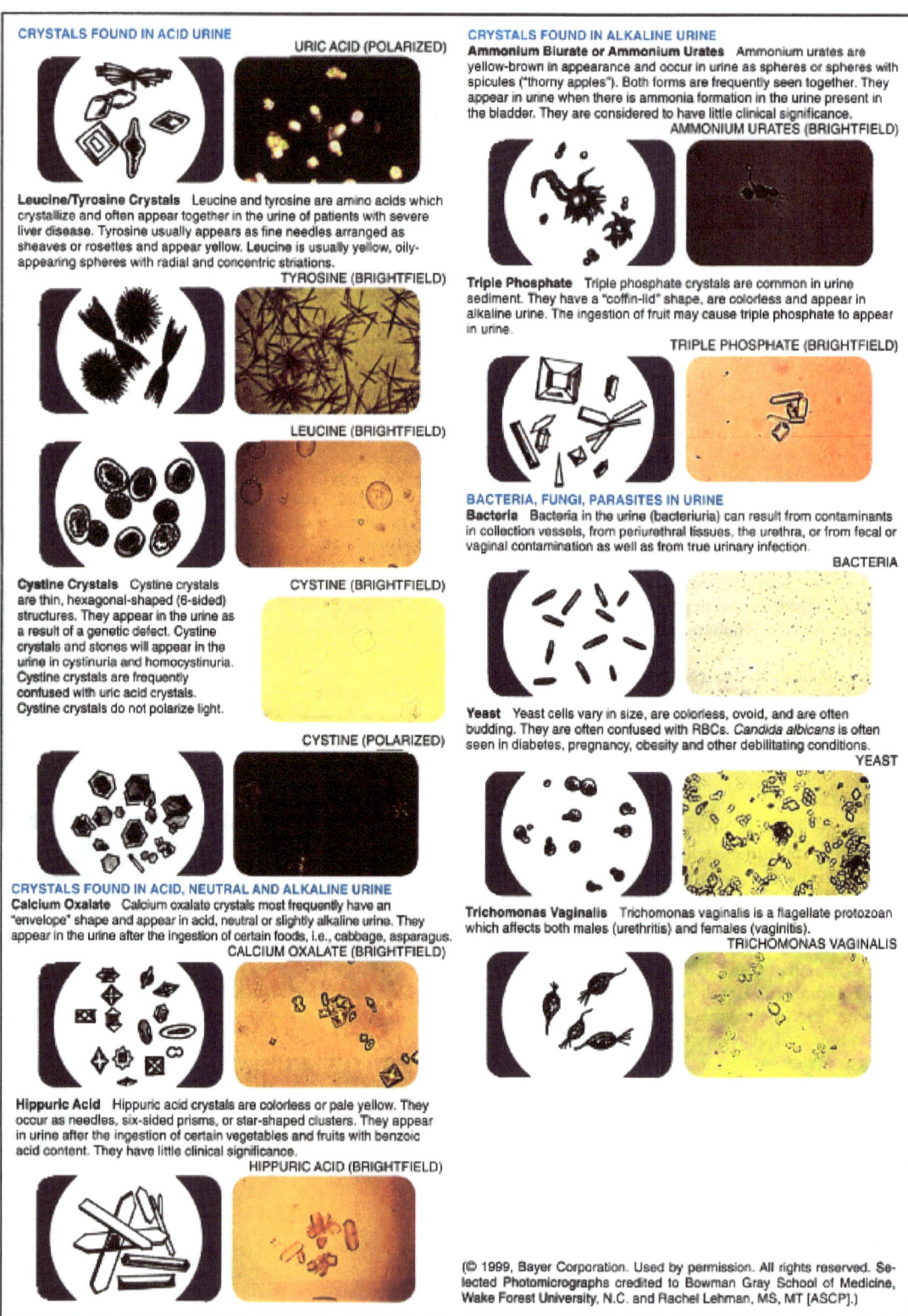

CRYSTALS FOUND IN ACID URINE

URIC ACID (POLARIZED)

Leucine/Tyrosine Crystals Leucine and tyrosine are amino acids which crystallize and often appear together in the urine of patients with severe liver disease. Tyrosine usually appears as fine needles arranged as sheaves or rosettes and appear yellow. Leucine is usually yellow, oily-appearing spheres with radial and concentric striations.

TYROSINE (BRIGHTFIELD)

LEUCINE (BRIGHTFIELD)

Cystine Crystals Cystine crystals are thin, hexagonal-shaped (6-sided) structures. They appear in the urine as a result of a genetic defect. Cystine crystals and stones will appear in the urine in cystinuria and homocystinuria. Cystine crystals are frequently confused with uric acid crystals. Cystine crystals do not polarize light.

CYSTINE (BRIGHTFIELD)

CYSTINE (POLARIZED)

CRYSTALS FOUND IN ACID, NEUTRAL AND ALKALINE URINE

Calcium Oxalate Calcium oxalate crystals most frequently have an "envelope" shape and appear in acid, neutral or slightly alkaline urine. They appear in the urine after the ingestion of certain foods, i.e., cabbage, asparagus.

CALCIUM OXALATE (BRIGHTFIELD)

Hippuric Acid Hippuric acid crystals are colorless or pale yellow. They occur as needles, six-sided prisms, or star-shaped clusters. They appear in urine after the ingestion of certain vegetables and fruits with benzoic acid content. They have little clinical significance.

HIPPURIC ACID (BRIGHTFIELD)

CRYSTALS FOUND IN ALKALINE URINE

Ammonium Biurate or Ammonium Urates Ammonium urates are yellow-brown in appearance and occur in urine as spheres or spheres with spicules ("thorny apples"). Both forms are frequently seen together. They appear in urine when there is ammonia formation in the urine present in the bladder. They are considered to have little clinical significance.

AMMONIUM URATES (BRIGHTFIELD)

Triple Phosphate Triple phosphate crystals are common in urine sediment. They have a "coffin-lid" shape, are colorless and appear in alkaline urine. The ingestion of fruit may cause triple phosphate to appear in urine.

TRIPLE PHOSPHATE (BRIGHTFIELD)

BACTERIA, FUNGI, PARASITES IN URINE

Bacteria Bacteria in the urine (bacteriuria) can result from contaminants in collection vessels, from periurethral tissues, the urethra, or from fecal or vaginal contamination as well as from true urinary infection.

BACTERIA

Yeast Yeast cells vary in size, are colorless, ovoid, and are often budding. They are often confused with RBCs. *Candida albicans* is often seen in diabetes, pregnancy, obesity and other debilitating conditions.

YEAST

Trichomonas Vaginalis Trichomonas vaginalis is a flagellate protozoan which affects both males (urethritis) and females (vaginitis).

TRICHOMONAS VAGINALIS

Figure 43-1. Continued